Energy 105

跳蚤和虱子
Fleas and Lice

Gunter Pauli

[比] 冈特·鲍利 著
[哥伦] 凯瑟琳娜·巴赫 绘
高 芳 李原原 译

上海远东出版社

丛书编委会

主　任：田成川

副主任：何家振　闫世东　林　玉

委　员：李原原　翟致信　靳增江　史国鹏　梁雅丽
　　　　任泽林　陈　卫　薛　梅　王　岢　郑循如
　　　　彭　勇　王梦雨

特别感谢以下热心人士对童书工作的支持：

匡志强　宋小华　解　东　厉　云　李　婧　庞英元
李　阳　刘　丹　冯家宝　熊彩虹　罗淑怡　旷　婉
杨　荣　刘学振　何圣霖　廖清州　谭燕宁　王　征
李　杰　韦小宏　欧　亮　陈强林　陈　果　寿颖慧
罗　佳　傅　俊　白永喆　戴　虹

目录

Contents

跳蚤和

虫子在一个人的头

发上相遇了。他们听说彼此很

长时间了，但还从未见过面。

"多美好的一天！"跳蚤打破沉默

说。"我是跳蚤家庭的父亲。你好啊！"

"你好，很高兴见到你！我一直期待

这一时刻。"虫子回应并自我介

绍："我是虫子家庭的母

亲。"

A flea and a
louse come across each
other in the hair growing on a
person's head. They have known
about each other for a long time but
have never met in person.

"Good day," says the flea, breaking the
silence. "I am the father of the flea family.
How do you do?"

"Hello, how lovely to meet you! I've
been waiting for this moment," says
the louse and then she introduces
herself: "I'm the mother of
the louse family."

跳蚤和虱子

A flea and a louse

我们享用着同样的生活空间

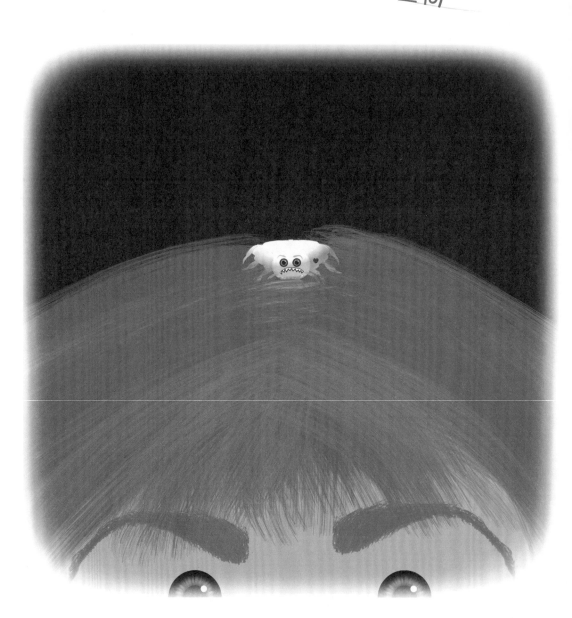

We both enjoy the same living space

6

"我一直在想我们
何时会见面，因为我们享用着
同样的生活空间。"
"的确如此，尤其是我们都爱咬寄主的
皮肤，如果可能，甚至吸点他的血！"
"我一直在想，为什么你们叫一个虱
子为louse，但两个或两个以上
的虱子为lice？"

"I've been wondering
when we would meet, as
we both enjoy the same living
space."
"Indeed, especially because we both
love to have a bite of the skin of our
host and, if possible, even a little bit
of his blood!"
"I've always wondered why you
call one louse a louse but
two or more lice?"

7

"这个我帮不了你，因为我也不知道。我倒是一直想知道，你是如何跳得那么高又那么远的？"

"实际上，我们没有跳。我们的膝盖上有个弹簧，释放的时候就像弓把箭射出去一样。可以说我们的身体结构更像一把弩。"

"你是世界上最好的跳跃者！"

"I can't help you there, as I have no clue. What I have been wondering is how you can jump so high and so far."

"Actually, we don't jump as such. We have springs in our knees that are released, like an arrow from a bow. You could say we are built like a crossbow."

"You are the best jumpers in the world!"

世界上最好的跳跃者!

Best jumpers in the world!

沫蝉才是真正的冠军

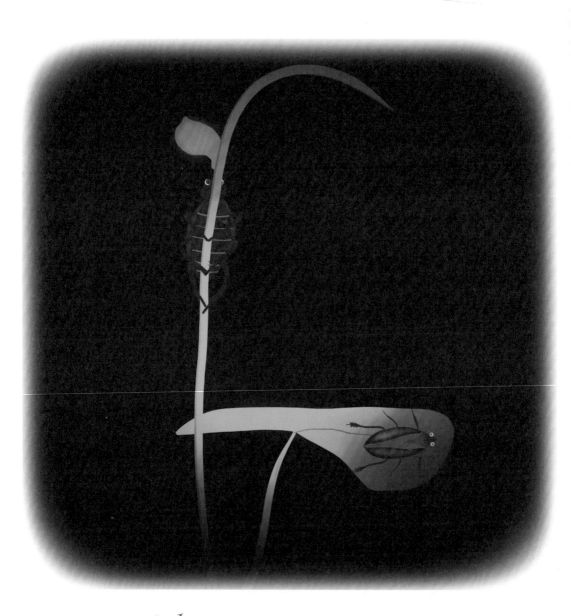

Real champion is the froghopper

"是的,
如果人类有我们的
弹跳力,他们就能向前飞跃
90米。但沫蝉才是真正的冠军,那
种小虫子可以产生相当于其体重400倍
的力量,而我们只能产生100倍,蚂蚱
甚至连10倍都不到。"

"这已经让人很佩服了。"

"Well yes, if
people had our jumping
power, they would be able to
leap 90 metres forward. But the
real champion is the froghopper.
That little bug can apply a force of
400 times its body weight, while we
can do only about a 100 times. A
grasshopper can't even do 10
times."

"That is still very
impressive."

"谢谢，但是你知道我们以动物为食，而沫蝉只吃植物。这可能是力量差异的原因。"

"我们也喜欢吃血液和皮肤，但我们却不能像你一样会跳。"

"或许是这样，但我听说你在数百万年前就已经寄居在黑猩猩和大猩猩身上了。"

"Thank you, but you know we feed on animals, whereas the froghopper only eats plants. That may be the reason for the difference in force."

"We prefer to eat blood and skin as well, but we cannot jump like you do."

"No, perhaps not, but I'm told that you were already living on chimpanzees and gorillas millions of years ago."

只吃植物

Only eat plants

人类甚至把我放进马戏团！

They have even put me in a circus!

"哦，是的，而且现在如果有机会，我们也会把人类作为食物的来源，无论他们喜欢与否。我想一下，我们在人类开始穿衣服时就寄生在他们身上了，衣服里面很容易躲藏，尤其是他们不常洗澡或者洗衣服的时候。"

"你不太受人欢迎，对吗？但我已经变得很受欢迎了，"跳蚤说，"人类甚至把我放进马戏团！"

"Oh yes, and now, if we get the chance, we will use humans as a food source too, whether they like it or not. I imagine we moved to people when early man started wearing clothing, and it was so much easier for us to hide in there, especially if they do not wash themselves or their clothes regularly."

"You are not popular with them, are you? I have become popular though," says the flea. "They have even put me in a circus!"

"我记得的！对于
我们俩来说，不幸的是，人类
发明了真空吸尘器，这使我们的生活
非常艰难。这就是如今在周围很少见到
你们的原因。"

"是的，这还不算什么。当空气太干燥
时，我们会生活得更艰难，"跳蚤
说，"那样我们就没有办法生
存下去了。"

"I remember
that! Unfortunately
for both of us, people have
invented the vacuum cleaner
and that makes life very difficult.
That is why there are now less of you
around."

"Yes, and that's not all. Life also
gets hard for us when the air is
too dry," says the flea.
"Then we have no way
to survive."

真空吸尘器

Vacuum cleaner

我们能够 "复活"

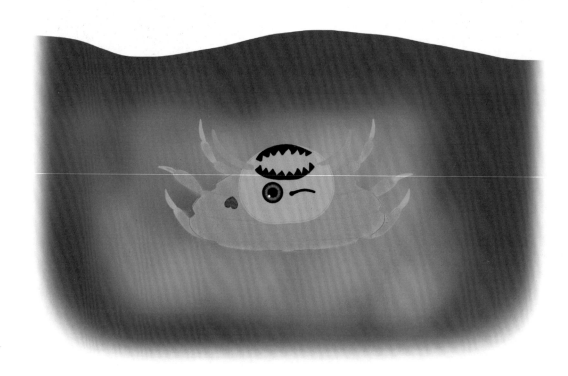

We are able to revive ourselves

"即使人类想用水淹，也无法摆脱我们。我们在水中12小时后，可能看起来像是死了，其实我们能够'复活'，生活依旧。"虱子说。

"我也不怕水，"跳蚤回答，"但我受不了他们把肥皂放入水里。那样我就失去了保护我不受细菌威胁的蜡，然后就会死掉。"

"Not even trying to drown us can get rid of us. After twelve hours under water, we may come out looking like death, but then we are able to revive ourselves and continue our lives," says the louse.

"I'm not afraid of water either," replies the flea. "But I can't stand it when they put soap in the water. Then I lose the wax that protects me so well from bacteria, and I die."

"我受不了我的寄主吃大蒜或者喝苹果醋。这让他闻起来很糟糕，我会失去食欲。这真的让我想换一个寄主。"

"对我来说很容易，我只需要把自己弹起来穿过房间就行，但可怜的你必须等到孩子们相互拥抱时才能摆脱那个散发着臭味的人！"

……这仅仅是开始！……

"And I can't stand it when my hosts eat garlic or drink apple cider vinegar. It makes them smell so bad that I lose my appetite. That really makes me want to find another host."

"It's easy for me to do that, as I just catapult myself across the room, but poor you, you have to wait until the kids are hugging each other before you can move away from a stinky one!"

… AND IT HAS ONLY JUST BEGUN!…

... AND IT HAS ONLY JUST BEGUN! ..

Fleas and lice are insects that have been feeding on people since the beginning of time.

自古以来，跳蚤和虱子就寄生在人类身上。

It is assumed that the body louse evolved from the head louse about 100 000 years ago, when people started to wear clothing and lost their body hair.

据推测，体虱大约在100 000年前由头虱进化而来，那时人类刚开始穿衣服并失去体毛。

虽子寄生在除了蝙蝠、鲸和海豚之外的所有鸟类和哺乳动物身上，甚至寄生在鱼类身上，比如养殖的鲑鱼。

Lice live on all birds and mammals except bats, whales, and dolphins and even infest fish like farmed salmon.

虽子孵化出来时就是父母的微缩版，称为若虫。跳蚤则是从卵孵化成幼虫，然后吐丝结茧并在其中化蛹，最后成为成虫。

Lice hatch as miniature versions of their parents, known as nymphs. Fleas hatch from eggs, develop into larvae, and weave cocoons with their silk, in which the pupae transform before they emerge as adult fleas.

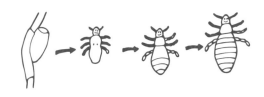

母兔临近分娩时，寄生在它身上的跳蚤会收到化学信号，开始产卵。当小兔子出生时，刚孵化的跳蚤就会转移到它们身上。

When a female rabbit is close to giving birth, fleas living on her receive a chemical trigger to start the production of eggs. As soon as the little rabbits (called kits, kittens or bunnies) are born, the newly hatched fleas will make their way towards them.

尽管人们讨厌跳蚤，跳蚤市场却是非常受欢迎的旅游景点。第一个跳蚤市场是1860年前后在巴黎建立的。当时贫民窟被拆除，商人们开始在街上售卖他们的商品。

Although people detest fleas, flea markets are very popular tourist attractions. The first flea market was held in Paris in around 1860, after merchants from the demolished slums started selling their wares on the street.

跳蚤可以被驯化和教导。自19世纪30年代以来，跳蚤马戏团在英格兰蓬勃发展。现在仅存的真正的跳蚤马戏团会在一年一度的慕尼黑啤酒节进行演出。

Fleas can be domesticated and taught tricks. The flea circus has thrived in England since the 1830s. The only remaining true flea circus now performs at the annual Oktoberfest in Munich.

跳蚤
马戏团

跳蚤表演时脖子上戴个小金项圈，与黄金皮带相连。在一些跳蚤马戏团中根本就没有跳蚤在表演，演员的工作是让人们相信确实是有跳蚤在表演。

Performing fleas wear tiny gold collars around their necks and are attached to gold wire leashes. In some flea circuses there are no fleas at all. It is the job of the performer to convince people that there are indeed fleas performing.

Think about It

想一想

If fleas and lice have been living along mammals for millions of years, will we ever find a way to get rid of them? Or should we welcome them as part of our family?

如果跳蚤和虱子已经和哺乳动物一起生活了数百万年，那么我们能否找到一个方法来摆脱它们呢？或者我们应该欢迎它们成为我们中的一员吗？

你是喜欢一个马戏团真的有跳蚤在表演，还是宁愿有一个小丑让你相信有跳蚤在表演呢？

Would you enjoy a circus where fleas are really performing tricks? Or would you rather have a clown make you believe that there is a flea performing?

How easy would it be to make a gold leash for a flea? Do you think it will ever get out of it alive?

为跳蚤制作一条黄金皮带容易吗？你认为它能活着摆脱这条皮带吗？

如果你用真空吸尘器清扫宠物睡觉的地方，你可以清理掉跳蚤和他们的卵，虱子和他们的幼虱。这样就足以摆脱这些寄生虫了吗？

If you vacuum the areas where your pets sleep, you will pick up fleas and their eggs, and lice and their nits and nymphs. Would that be enough to get rid of these parasites?

Let's put together a
flea remedy that smells pleasant
and keeps pests at bay.
You will need the following:
A sachet made from a natural and breathable fabric.
(You can use an old scarf.)
A handful of cedar chips.
Two teaspoons of dried lavender buds.
The peel of one lemon.
Place the cedar chips, lavender buds, and lemon peel in the
sachet or scarf and sew it up. Place the bag next to where
you pet sleeps. This will keep your pet (and you)
free from fleas!
You will need to refresh the content
every two months.

让我们一起做一个跳蚤治疗袋吧，它气味宜人，还能让害虫不敢靠近。你需要以下物品：一个透气的天然织物袋（或一条旧围巾），少量的雪松碎片，两勺干薰衣草花蕾，柠檬皮。把雪松碎片、薰衣草花蕾和柠檬皮装入袋子或围巾中缝起来。把它放在你的宠物睡觉地方的旁边。这会让你和你的宠物摆脱跳蚤！你需要每2个月更新一次里面的东西。

学科知识
Academic Knowledge

生物学	寄生虫对健康的影响；头发对身体的作用；寄生虫摄取的营养来自血液和死亡的体细胞；不完全变态昆虫（如虱子）的若虫对应于完全变态昆虫（如跳蚤）的幼虫和蛹；植食性动物和肉食性动物的区别；跳蚤有坚硬的外壳。
化 学	使用化学物质杀死寄生虫；肥皂能溶解蜡和脂质；大蒜中的化学物质能够产生一种强烈的气味；如何制造醋。
物 理	弹簧弹力与肌肉力量的区别；肥皂的物理效应；亲水蜡和防护涂料的功能；干燥剂。
工程学	美洲和非洲原住民使用的弓和箭的功能；弓和弩在技术、性能上的区别。
经济学	成本因素：寄生虫造成生产力损失，以及化学品污染和后续健康问题带来的成本。
伦理学	奴役动物以娱乐人类所带来的伦理问题；以谦虚的态度接受如下事实：总有某人在某一点上比你优秀。
历 史	寄生虫的进化：从寄生在大猩猩和黑猩猩身上到寄生在人类身上。
地 理	跳蚤和虱子在世界各地随处可见。
数 学	如何计算你从一个地方跳跃到另一个地方所需的体力；如何计算克服重力所做的功。
生活方式	服装的产生；人体卫生、洗衣服和洗澡的重要性；在家里使用吸尘器对卫生和健康的影响；人们第一次见面时问候形式的差异；从古希腊时代起，薰衣草就被用作驱虫剂。
社会学	对气味的不同感知和欣赏：一个人觉得愉悦的气味可能另一个人却排斥；当孩子们玩耍时，他们彼此频繁接触，这使寄生虫在他们之间转移比在成年人之间更容易。
心理学	自我认可的重要性；如何表达热情；如何询问你不懂的东西，尽管那会显得你很无知。
系统论	物种如何为适应新环境而演化、变异和转化：一些物种，例如被认为是害虫的跳蚤和虱子，如何仍能成为生态系统的组成部分，并在其中找到自己的容身之地，即使人类还未能很好地了解其功能。

情感智慧
Emotional Intelligence

跳 蚤

跳蚤礼貌恭敬地向虱子自我介绍。他很体贴地问一些问题以便更好地理解虱子，并与她建立友好关系。他还饶有兴趣地倾听虱子的问题，并花时间来解释她不理解的事情。即便当虱子表示对他良好弹跳力的尊重时，他还谦虚地告诉她有比自己做得更好的，并解释了为什么会出现这种情况。跳蚤在接受虱子赞美后表示友好，并试图指出虱子的特点。跳蚤了解自己，知道自己因掌握马戏表演技巧而受欢迎。当虱子突出表现自己的独特优势（能在水下生存超过12小时）时，他知道自己也能做到，但并没有夸耀自己，而是分享了恐惧。跳蚤很同情虱子，因为他很容易改变寄主，而虱子这样做却需要更多的耐心。

虱 子

虱子在介绍性问候时不是很正式，并且首先分享她和跳蚤的共同点，也就是他们的寄主相同。她谦虚地承认不是一切都知道。她通过提问题试图寻找共同点，表现出钦佩和尊重。当跳蚤希望突出虱子的一个特点时，她发现好像没有什么是与众不同的。跳蚤谈到自己的声望时，虱子温柔地告诉他，吸尘器的发明是周围跳蚤数量减少的原因。然后虱子指出自己的一个优点和一个弱点，使这两只昆虫产生了共鸣。

艺术
The Arts

你要去参加一个脱口秀节目。站起来，假装有跳蚤在你的手上。让跳蚤跳跃、翻筋斗、再翻一个筋斗。让所有观众相信真的有跳蚤在你的手上。假装跳蚤跳到第一排的一位女士身上——也许是你的老师，或假装在她的头发里"找到"一只跳蚤。扮演虚拟的跳蚤马戏团主持人是很有趣的。

思维拓展
Systems: Making the Connections

人类对跳蚤、虱子以及其他害虫的反应是试图消灭它们。然而，自人类存在开始（甚至在此之前），跳蚤和虱子就已经存在了。数千年后，问题不在于我们能否找到一种新的化学物质最终消除所有害虫，而是如果我们可以设计自己的栖息地，跳蚤和虱子就无法在其中找到容身之地了。尽管我们可能还没有认识到虱子和跳蚤的生态价值，但必须承认，摧毁我们不喜欢或不理解的事物可能会影响社会的可持续发展。如果虱子和跳蚤让我们不舒服，那么我们应该改变它们的生存条件。首先定期用真空吸尘器清理房子，特别是在狗和猫睡觉的地方。然后，不要把跳蚤、虱子及其幼虫和卵放置在垃圾箱中，它们会繁殖，所以要立即处置掉，确保它们不会再次进入房子。接下来，为这些寄生虫创造不适条件，在我们的日常饮食中添加大蒜和苹果醋。或者你可以使用植物，它们的气味对人类来说是愉悦的，而跳蚤和虱子却很讨厌。自古希腊和古罗马时代以来，薰衣草就被用来控制跳蚤和虱子，它在阿拉伯文化中也很普及。这个根除和排除的区别类似国际象棋和围棋的区别：国际象棋游戏的目的是把王将死，而围棋游戏的目的是让对手无处可去。

动手能力
Capacity to Implement

在网上或报纸上看一看，你会发现如果学校有跳蚤和虱子爆发，学校会被迫关闭一段时间。找出当局应对有关问题的措施。提出一个你能想到的策略，首先要扭转危机，其次要避免问题再次发生。现在列出解决问题所需的成本，对如何使用这些钱，你是否有建议？描绘一下，如果采用了不同的方法，新的业务活动应该怎样开展。

故事灵感来自
This Fable Is Inspired by

蒂姆·科克里尔
Tim Cockerill

蒂姆·科克里尔是一名昆虫学家，也是一个马戏团演员，毕业于英国剑桥大学，获博士学位。他专门从事生物多样性和生态系统科学研究，研究生命之间的相互作用。他攻读动物学博士学位期间关注研究婆罗洲的热带雨林，那是地球上生物多样性最丰富的地区之一。由于棕榈种植园的栽培，这些热带雨林正在经受威胁。在几个月的野外研究中，他发现了大量的新昆虫物种，现在这些物种已经被永久收录在伦敦自然历史博物馆的《昆虫学集》中。蒂姆也是一个经验丰富的演员。他是一个吞火魔术师，也是一位研究跳蚤马戏团的专家。

图书在版编目(CIP)数据

冈特生态童书.第三辑修订版:全36册:汉英对照 /
(比)冈特·鲍利著;(哥伦)凯瑟琳娜·巴赫绘;
何家振等译.—上海:上海远东出版社,2022
书名原文:Gunter's Fables
ISBN 978-7-5476-1850-9

Ⅰ.①冈… Ⅱ.①冈… ②凯… ③何… Ⅲ.①生态环
境–环境保护–儿童读物—汉、英 Ⅳ.①X171.1-49

中国版本图书馆CIP数据核字(2022)第163904号
著作权合同登记号图字09-2022-0637号

策　　划　张　蓉
责任编辑　祁东城
封面设计　魏　来　李　廉

冈特生态童书
跳蚤和虱子
[比]冈特·鲍利　著
[哥伦]凯瑟琳娜·巴赫　绘
高　芳　李原原　译

记得要和身边的小朋友分享环保知识哦！
八喜冰淇淋祝你成为环保小使者！